THE GENERAL THEORY OF RELATIVITY

BY ALBERT EINSTEIN

SIMPLIFIED FOR KIDS

SAMUEL KINYANJUI

Contents

1. The Beginning 3
2. Consequences of The Theory of Relativity 4
   - 2.1 The View of Time 4
   - 2.2 Space- Time Concept 5
   - 2.3 Length Difference/ Dilation 6
   - 2.4 A New View of Gravity 7
   - 2.5 Gravitational Lensing 8
   - 2.6 Distortion of Planet Mercury's Path 9
3. Challenges faced in the acceptance of the theory 10
4. Scientific Proof of the General Theory of Relativity 11
5. Modern Proofs and Evidence 14
6. Changes the Theory had to the Modern World 16
   - 6.1 Black Holes 16
7. Applications of the Theory 19
   - 7.1 Global Positioning System (GPS) 19
   - 7.2 Electromagnets, Transformers and Electrical Generators ... 20
   - 7.3 CRT Monitors 21
   - 7.4 Nuclear Power Plants and other Nuclear Applications 22
8. Conclusion 23

Works Cited 24

1. The Beginning

Albert Einstein pioneered the thought that the laws of physics appear the same to everyone. He also pioneered calculation that the speed of light (299,338 kilometers per second) is always constant everywhere in the universe. This was contrary to the beliefs held by his predecessors who deemed that the speed changes due to the existence of a luminiferous ether, depending on the relative motion of the source of light and the observer.

From these deductions Einstein came up with the foundations of the Relativistic Approach of viewing the universe. There is no fixed reference point of viewing events but everything moves relatively to everything else in the universe. This was later coined as the Special Theory of Relativity. The term

Special is as a result that Einstein first viewed events to be moving/ happening at a constant speed and not accelerating or decelerating. From this Einstein discovered some weaknesses in that the theory was limited to constant speeds only. However, there is more to that in the universe acceleration of moving bodies must be considered and thus the advent of the General Theory of Relativity which considered events accelerating or decelerating relative to each other. (Schwinger).

2. Consequences of The Theory of Relativity

2.1 The View of Time

The traditional aspect of time was that time moved/ ticked at the same rate for everyone.

However, this is not the case as time or a clock ticks slowly for a person relative in motion as compared to the time of a stationary observer. This is called Time

Dilation. As speed of the moving observer continues to increase relative to the stationary observer the time ticks continuously slower to almost utmost standstill if the traveler nears the speed of light. Time and space are connected in a mesh like fabric called space-time. It is tied up to the other space three dimensions, as a fourth dimension. Thus, it's equally susceptible to changes caused by gravity, relative motion and those occurring on space itself.

2.2 Space- Time Concept

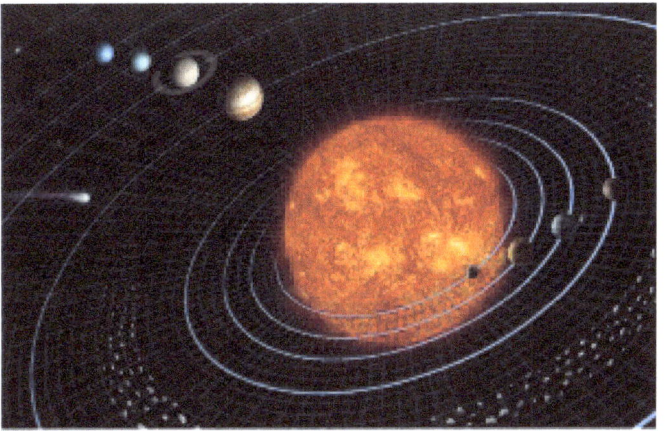

As observed above Einstein discovered time did not flow 'alone' unbound by other factors but co-existed with the other space dimensions of length, width and breadth. Thus, time is viewed as the fourth dimension and hence the concept of Space-Time. This is primarily because Time does not flow independently without the influence of other factors in the universe. It is tied up to the other space three dimensions, as a fourth dimension. Thus, it's equally susceptible to changes caused by gravity, relative motion and those occurring on space itself.

2.3 Length Difference/ Dilation

Due to relativity the moving objects seem shorter in length along the motion frame as compared to when they are stationary. This change is very minute to register and becomes much more profound when the moving object nears the speed of light. This phenomenon is called Length Dilation. Under ordinary day to day speeds such contractions appear are negligible. The effect unfortunately is very subtle, however in science a change regardless of how minute is always a change and worth noting. For example, if a body say a car theoretically was to move at 13400 kilometers per second, its contracted length would only be 0.99 of the original length, that is has shrunk by only 1%. (Pyykkö)

The formula governing the length of the object as speed increases is simply:

$$L = \frac{L_o}{\gamma(v)} = L_o\sqrt{1 - {v^2}/{c^2}}$$

Eqn 1. Length Dilation Equation

Where:

L= the Length as seen by the stationary observer

L_o = the normal/proper length of the object from rest position

$Y(v)$ = the Lorentz Factor Defined as $\gamma(v) = \frac{1}{\sqrt{1-v^2/c^2}}$

V = relative speed as observed relative to the moving object

C = speed of light

This again shows that Einstein was right when he postulated that events occur relatively to everything else and even length or spatial dimensions changed in relative motion.

2.4 A New View of Gravity

From this theory Einstein attempted to include Gravity in it. From Sir Isaac Newton's time, Science had it that Gravity pulls things to the ground, and it was caused by the mass and speed of objects e.g. The Earth's Rotation and hence its gravity pulling you downwards, but to Einstein this was not right. To him Space-Time was a 'fabric' that was 'distorted' by the momentum and energy of objects as they moved through Space-Time so as to cause what is called the "pushing effect ".

What does this mean? Basically, the effect says that gravity warps/ distorts Space-Time.

Take an example of a coin in a glass of honey, rotating the coin makes the honey surrounding it warp and thus swirl around in the direction of the coin. Similarly, Space-Time is 'swirled' by massive objects such as the earth, stars and the sun, the same effect keeps the earth and other planetary objects in revolution around the sun.

Space-Time is everywhere in the universe this fabric exists even in the air surrounding us. Earth does not pull things downwards, what actually happens is Earth's momentum and energy (due to its rotation and revolution) causes Space-Time to distend or be distorted and thus due to that distortion, Space-Time pushes you downwards towards the ground and thus the apparent feeling of Gravitational Pull.

2.5 Gravitational Lensing

This is one of the effects where gravity is factored in the Special theory of relativity. This improved the theory as now Einstein could extend it to objects of accelerating nature relative to one another not just moving at constant speeds.

Gravitational Lensing is a key effect and is the main experiment that Einstein took to prove his theory right. We will embark on the experiment later on.

As previously discussed, it is evident that Space-Time is distorted significantly by gravity of massive objects such as the sun. Light from distant stars beyond the sun have light traveling from them bent/ distorted as it approaches earth. This is because the path (space-time) is bent around the sun and thus since light travels in a straight line it only means that its path is bent. This is a significant discovery that reinforced the theory (Atek).

2.6 Distortion of Planet Mercury's Path

Einstein postulated that the orbit of the planet mercury was not a regular orb as accustomed to other planets. This was because it changed frequently as it revolved around the sun. Some scientists even thought that due to the closeness of the planet to the sun the two might be in a tidally locked system. This is the system that exists between the earth and the moon, where the moon revolves around the earth with only one side facing the earth and the other side (dark side of the moon) always away from the earth. However, for mercury and the sun this was not the case and in fact the planet rotates thrice for every two revolutions.

Thus, the real issue with mercury is that the sun being so massive and being so near warps space-time greatly around the normal orbit of mercury and thus distorts the planet's path as it makes its way round the sun. This is another effect of the theory of general relativity.

3. Challenges faced in the acceptance of the theory

Einstein view of the universe, time and gravity seemed to take all the classical laws of physics and topple then, upside down. This meant that he had to defy some of the greatest minds that have postulated theories on gravity and even our personal view on the same. This led him to consult with Max Planck, the only scientist in time who understood what Albert was talking about. He advised him that if he chose the gravity way it would mean going against Sir Isaac Newton's 17th century discoveries. This was difficult for Einstein and the whole scientific community. He presented his paper to the German Scientific congress at Berlin, but it was greeted by low understandability and skepticism. Einstein discovered that he needed to find scientific proof to validate the Equation of General relativity within which the theory was based.

4. Scientific Proof of the General Theory of Relativity

The only way for Einstein to validate the theory was conduct the experiment that was simple enough to understand but also effective enough. It is common knowledge that light travels in a straight path/line. However, beams of light at times bend in their travel. Why? The only explanation is that the path must be bent and hence light appears distorted.

What is big and massive enough to curve space so that light traveling close it seems to bend? The Sun was Einstein's answer. This was the deal. The light from stars beyond the sun would be curved as they approached the sun. Then as the light reached the earth the stars would seem shifted from their normal position as an optical illusion. This is Gravitational Lensing. The challenge however is, it is impossible to view stars at day time due to the sun's brightness, unless in a total Solar Eclipse.

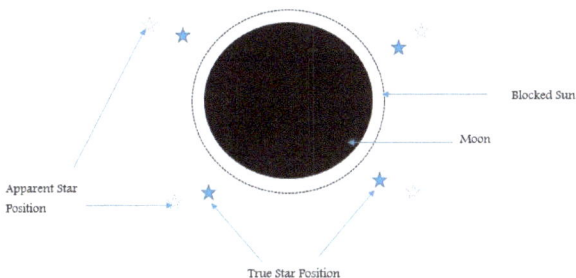

Fig 1. The Gravitational Lensing on Starlight as seen from Earth during a Solar Eclipse

Thus, the gravitational lensing presents an optical illusion to the position of the stars due to the bending of space fabric by the sun's gravity and thus the apparent bending of light.

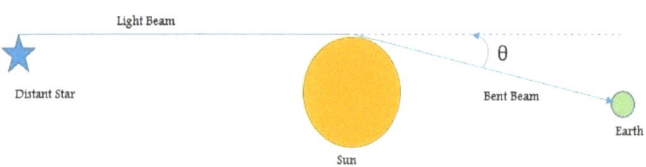

Fig 2. The illustration of the bending of light due to the Sun's gravity.

The diagram shows how gravitational lensing occurs. The angle θ represents the angle through which the light is bent. Einstein first calculated this angle as θ = 0.84" however later he discovered that he had been erroneous in his calculations and thus discovered it was about twice the initial prediction thus θ = 1.75". This correction was sparked by a moment of brilliance when he reconsidered planet Mercury's distorted orbit as explained earlier under the Effects of the Theory.

Einstein thus knew astronomers/ astrophysicists were only people to photograph the stars during the eclipse. Einstein contacted Erwin an astrophysicist in Berlin and William Campbell of Lake Observatory California to photograph the eclipse at Crimea Russia on 21st August 1914. The experiment was not successfully as the clouds blocked the eclipse. The astrophysicists' equipment was confiscated by the Russians on the onset of World War II.

The next solar eclipse occurred in Washington State on 8th June 1918. Astrophysicist Campbell took this as a new opportunity and managed to have some shots of the stars during the eclipse. The photographs were reviewed to determine if the gravitational lensing had occurred. It was estimated the θ= 1.75" would yield about 1mm deviation in the stars' position. However, Campbell did not get such results and almost discredited Einstein's theory. Before this could happen, an English astrophysicist in the name of Arthur Stanley took an expedition to the island of Principe in Africa on 29th May 1919. Another eclipse was clearly observed there. From the results, Arthur concluded that Einstein might be right.

The collision between the two astrophysicists' results warranted for a final expedition. The next total solar eclipse was to be observed September 1922. Both Arthur and Campbell were present, among other observers from Australia and India. This time both consented that 92 stars observed seem all shifted from their normal position and by the estimated distance of θ = 1.75".

> Finally, this was the proof that Einstein and the whole world had waited for.

5. Modern Proofs and Evidence

The scientists nowadays do not have to wait till the solar eclipses can reveal Einstein's Theory. NASA recently published a report, crediting the Theory, "Einstein Proved Right Again" rung the news. This was due to the recording of Gravitational Ripples that occurred when two massive Blackholes collided with each other.

Black holes are the dense remains of stars after the fuel powering them (nuclear fusion) has died out. Black holes are extremely dense to a point that their huge gravity can never allow anything to escape even light itself, and thus their name.

The ripples are a further proof of the existence of the Spacetime fabric. The two blackholes were massive enough and have enough gravity to send distention ripples and warps on spacetime across the universe detectible on earth. This happens in the same way two ships would collide at sea and the ripples are set forth on the water and this is synonymous to the Gravitational Ripples observed by NASA on 11th Feb 2016.

The two black holes had each a diameter of 50km and one was the equivalent of 30 suns in mass while the other had around 26 solar masses. These masses are extremely huge despite the small sizes of the blackholes.

6. Changes the Theory had to the Modern World

6.1 Black Holes

A lot of the understanding today on black holes is credited to the analysis of the General Theory of Relativity. Before then, they were mysterious if even known to exist. One aspect that puzzled scientists for example, was how matter behaved upon reaching the Event Horizon.

They discovered that the laws of physics and indeed the laws of nature collapse upon entry into a black hole. If an astronaut was to fall inside one, we would observe him falling, then as he reached the Event Horizon, we would see him fade out or as many would say disappear. This is because time freezes in the Event Horizon, and you can no longer observe the happenings. This Time Dilation is credited to Einstein's theory due to the great warping of spacetime around the dense Event Horizon. (Hofmann).

6.2 Worm holes

Worm holes are to some degree the opposites of black holes. When a black hole would suck you to oblivion, a worm hole may create a space channel or path to another location in the universe. Take this analogy as an explanation:

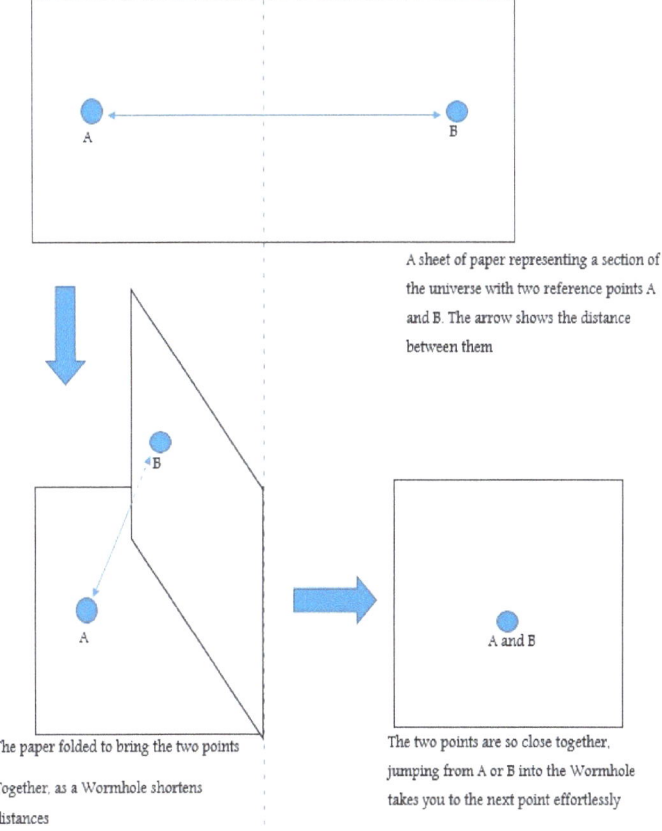

Fig 3. The Analogy on a Sheet of Paper to describe Wormholes

A sheet of paper represents the section of the universe with two points A and B. The shortest distance at first is the straight line in between them. The wormhole behaves as indicated and warps spacetime fabric like a sheet of paper. This is because of the huge gravity it possesses. The distortion/folding shortens the distance between the two points and thus makes it possible to travel through vast distances in space. (Bronnikov).

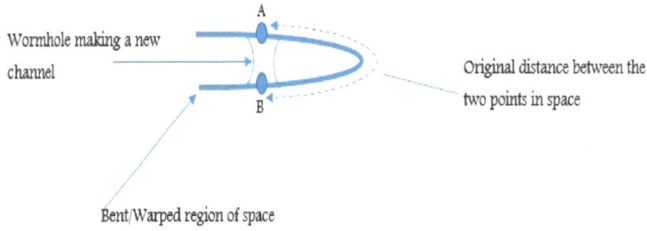

Fig 4. A better understanding of the wormholes.

The understanding of how wormholes behave is credited to Einstein's theory of General Relativity.

7. Applications of the Theory

7.1 Global Positioning System (GPS)

This is an important tool in navigation as nowadays even vehicles are fitted with GPS devices to tell its location anywhere on the earth. This is only possible by the use of satellites that factor relativity during their operation. Satellites orbiting the earth are doing so in a very high speed so as to not to fall back to the earth and thus the relativistic effect must be considered. To cater for time fluctuations and dilation, they use that are extremely accurate (a few nanoseconds accuracy)

If ignored the effect would be large if left unchecked the speed of the satellites which is over 6000 mph and the gravity causes a time dilation of 7 microseconds per day. Translating to 0.02555 seconds per decade. Most satellites serve for more than that period. Thus, the spot-on accuracy provided by GPS would be lost gradually.

7.2 Electromagnets, Transformers and Electrical Generators

According to Faraday's Law, the change in magnetic flux linkage results to increase in current flow. If you move a wire across the magnetic field, say of permanent magnets, there is flow of current in the wire to relative flow of electrons in the conductor to the magnetic field. However, if the wire was now stationary but the magnets rotated, the current is still induced in the conductor. This shows there is no absolute reference of observance of the charged particles, as stated by Einstein's theory. The rotation and electricity production are the basis of the functioning of generators and thus the production of electricity.

7.3 CRT Monitors

Cathode ray tubes such as the old television sets are also an application of relativity. The tvs contain the electron gun that shoots a beam of electronics to the screen. The beam of electrons contains light photons which each represent a pixel upon striking the phosphorescent screen. These photons are moving at nearly a third of the speed of light. The relativistic effects such as length contraction had to be taken into consideration when designing the cathode ray tubes, and in particular the electron gun. (Hainich).

7.4 Nuclear Power Plants and other Nuclear Applications

Decaying radioactive isotopes such as Uranium rods are used as the fuel in nuclear power generating plants. The regulated/controlled decay is the breakdown of the isotopes into other elements and thereby emitting heat used to heat water and the steam to turn electrical turbines. Relativity is the only reason and explanation as to how mass and energy can be interconverted from one to another.

In a nuclear weapon the decay of radioactive isotopes is also the principle of operation, however in this case the release in energy is not controlled. The decay is exponential and unregulated and thus the explosion occurs, due to the huge amount of energy released at once. This is testament to Einstein's equation: $E=mc^2$

Eqn. 2 The relationship between mass and energy

Where E= Energy, M=mass, C= Speed of light

8. Conclusion

Einstein's Theory of General Relativity, is a milestone for scientific discoveries and future achievements. Very few advancements to the theory have been made since its inception and it remains proven by modern scientific methods. It has also led to special applications and a better understanding of space-time, matter, energy, celestial mechanics and the entire universe. This is one of the most influential and beneficial discoveries of the 20th Century, long live Albert Einstein.

Works Cited

Atek, Hakim, et al. ". "New Constraints On The Faint End Of The Uv Luminosity Function At Z~ 7-8

    Using The Gravitational Lensing Of The Hubble Frontier Fields Cluster

    A2744." The Astrophysical Journal 800.1 (2015): 18.

Bronnikov, K. A., & Krechet, V. G. "Rotating cylindrical wormholes and energy conditions."

    International Journal of Modern Physics A, 31(02n03), 1641022. (2016).

Hainich, Rolf R., and Oliver Bimber. Displays: fundamentals and applications. CRC press,, 2014.

Hofmann, S., & Rug, T. ". A quantum bound-state description of black holes." Nuclear Physics B, 902,

    (2016): 302-325.

Pyykkö, Pekka. ""Relativity, Gold, Closed-Shell Interactions, and CsAu· NH3." Angewandte Chemie

    International Edition 41.19 (2002): 3573-3578.

Schwinger, Julian. "On angular momentum." 2015.

www.ingramcontent.com/pod-product-compliance
Lightning Source LLC
Chambersburg PA
CBHW040351220526
45473CB00009B/2856